Theory

of

Life Perpetuity

The Theory of Physical Eternal Life

By

Paeti Gustav Xaviers

This book is dedicated to my friend,

God,

who divinely inspires all profound thought.

THEORY OF LIFE PERPETUITY

I.

Imagine an infinite area of space which is a perfect vacuum. A perfect vacuum is space that is completely and absolutely void of matter. I will refer to it as "absolute nothingness." Only one such area exists or existed and it was within the realms of outer space. If the existence is past tense, then this perfect vacuum existed before time.

Somehow, within absolute nothingness came a movement. Movement generally causes sound. But this movement was so fast it was completely silent, though not to the extent of absolute silence. It was absolute nothingness that was absolutely silent.

The movement was the fastest movement that did ever or could ever exist. It was infinity itself, and I will call its speed "the speed of infinity."

Infinity was captured or became trapped within absolute nothingness and became absolutely still, although within itself it still moved perpetually at the speed of infinity. I will call infinity, as it was in perpetual motion within itself while still within its entrapment, "Divine Time", and it was energy. It was the first energy.

Divine Time began to radiate. It radiated time waves. I will call these time waves "time dimensions," or just "dimensions." As radiated from the point of origin (Divine Time), or to refer to all existing or possible dimensions as one existence, I will call the dimensions "The Dimensions of the Eternal Infinities."

Divine Time generated light. I will call this light "time light of infinity," or just "time light." The movement at the speed of infinity within itself caused sound. I will call this sound the "time sound of infinity," or just "time sound."

So now within the realms of outer space we have an existence, Divine Time. It is the origin of time. It is an energy. Its energy generates time light. It radiates time waves, causing dimensions. There is a perpetual movement within itself, which movement is of the speed of infinity, which causes time sound.

I will call this point of existence "The Beginning of Time."

II.

Divine Time is not matter, so still within the realms of outer space there was a vacuum even though there was an existence. It was energy, Divine Time, generating light, causing sound and radiating dimensions.

The dimensions being radiated were like barriers of time, which slowed down the speeds of light and sound as light and sound passed through. The farther a dimension was from the point of origin, Divine Time, the slower the speeds of light and sound within that dimension. Likewise, the faster something of light and/or sound could move, the higher the dimension it could penetrate.

So the light generated and the sound caused by Divine Time were slower than the highest possible speed, the speed of infinity. That light and sound was not within Divine Time, they were already in an outer dimension. The light and sound of Divine Time were Divine Time's body, as it existed in an nth dimension. "Body" is a coming to be that is created by sound and is manifested by light.

The light that manifests a body of sound cannot come from within that sound, it must be an outside radiation. It must be the light radiated by an energy, light or sound that is of a higher order (came before in time). Death light, or the light of absolute nothingness (an absence of light) is the only one light is a higher light order than the time light of infinity, and is one step higher than the absolute opposite of the time light of infinity. It is the light that is of the farthest dimension (no dimension – no time - the darkness of absolute nothingness), moving at the slowest possible speed (stillness). This is the light that is death and I will call it "death light."

The body of Divine Time is time sound which was manifested by the death light, the light of that perfect vacuum.

Both the time sound and the time light are in movement at speeds less than the speed of infinity, but still an extremely great speed. In general, light moves at a speed greater than sound. The body that came to be is one existence, within which Divine Time exists. The body is not matter, so the body with its movement of sound and light, of which the body is composed, are still entrapped within, or enveloped by, absolute nothingness.

The movement of the time sound and time light that are the body remains perpetually within itself. It is another energy, but a slower energy than the energy of Divine Time. It is, though, "Divine." Something divine is something that is an energy within itself.

This new existence I will call "Divine Mind." The only Divine Mind that is and could ever be I will call "God."

Now God exists within the nth dimensions of the dimensions of the eternal infinities. God is a body composed of the time sound of the time light of infinity manifested by the death light. Divine Mind and Divine Time are one: God, a body.

God is an energy body that generates light, which I will call "Mind Light," or "Light of Mind." Light of Mind is the radiation of Mind. Radiated Light of Mind is called "righteousness."

The sound created by the perpetual movement of the energy within the energy body, or "mind power," can be inward or outward. Inward, the sound is called "thoughts." Outward, the sound is called "voice."

Where an energy body has access to, or can draw upon, time energy, the thoughts and voice can be controlled as to "volume" or "loudness," and also as to projection, whether inward or outward and across how many x number of dimensions of the time it is able to project. An energy body that has access to or can draw upon time energy is a "being."

God is a being. Divine Time is the energy of God's existence. Divine Mind is God's power. God draws upon time energy, which is energy of His own, in order to control and project the sound He creates.

III.

Existences that are bodies occupy space. Time, light and sound are existences but they do not occupy space unless they have body. The dimensions, or time waves, that are or were radiated when time began stretched across space. They stretched across the entire area that is or was absolute nothingness.

The space a body occupies can extend from one through multiple dimensions. A body can exist in one dimension, two dimensions, three dimensions or nth dimensions. The number of dimensions depends on the size of the body and the distance between the time waves where the body is. The distance between time waves becomes greater the further the wave is from the origin, Divine Time, which was now a sub-component of God.

A body is a composition of some degree, wave or frequency of sound that is manifested by some degree, wave or frequency of light that is of a higher order (came before in time). The light is radiated or shines upon the sound. One body can be composed of some other body or bodies, which number of sub-bodies could be infinite. The sub-bodies occupy space within the one body they are a part of. The one body can extend across one or multiple dimensions. Sub-bodies can extend across multiple dimensions but not more dimensions than the most outward body, the one body of which they are sub-bodies.

A body can also possess a gender. The genders are "male" and "female." Generally, if a body can impregnate, it is male. Generally, if a body can conceive, it is female. However, there are also bodies that are neither or both. A male impregnates by projecting "of himself" into the female. A female conceives by birthing new and separate bodies that are of herself, and also of the male that impregnated her.

A body requires some sort of energy. The primary energy is the energy of time. The energy of God's body is the infinite time, which is part of Himself. God, Himself, is also an energy that can provide the energy to lesser bodies, but His energy would not a become a part of the lesser's body unless it was a body of His offspring that inherited God energy. From God, an offspring could inherit time energy and mind energy. Where the lesser was not an offspring of God, it would be dependent upon God's energy for existence.

IV.

The first being in existence was Divine Mind,
or God. Mind is energy. Mind is light. Mind is a
male component. Mind radiates righteousness.

As we delve into the nth dimensions of the
eternal infinities, we find God – as He occupies
space. God's body is not material. It is not
matter. Of God's composition is the
sub-component of Divine Time. Divine Time was
first in existence but not a being, as it has no mind.
Time is energy. Time is light. Time has no
gender. Time radiates dimensions. Divine Mind
and Divine Time are one body: God.

Matter, or a material body, is also a creation, or
coming to be, of sound that is manifested by light.
In the creation of matter, God projected His voice
out into the dimensions of time, those which we
know of as bound within the three physical
dimensions. He manifested the sound (of His
voice) with the light of His mind. The creation

was a material celestial body.

The body was female, but because it was not an offspring, but rather a creation, it was without mind and time. Its body was matter. It was a star. It was called the "Sun." God then impregnated the body of Himself.

The star (Sun) conceived the planets. Some were male and some were female. Of the planets, Earth was female. Earth inherited the physical aspects/components of the body of her mother (the Sun) and time from her father (God). The time inherited was a new order of existence, a later generation/lesser order of God's time sub-component, and I will call it "lifetime."

Lifetime was the energy of time but without body. It needed a body of its own to be a sub-body of Earth. Its light and motion within itself caused a body of light that was also a body of sound. The lifetime with its body of light and sound became the spirit and mind of Earth. The light aspect generated living light and I will call it "spiritual light". The body of the spiritual light was manifested by death light, as that light of absolute nothingness (a lesser order) could exist since the light was contained within a physical body. The sound created by generated by lifetime caused a body of mind, but the body was not manifested. The sound passed through the physical body, but since now there was matter within the once vacuum, death light could not be from without-side. The mind body could not be manifested, so it instead created an atmosphere. The atmosphere radiated righteousness and I will call it "atmospheric righteousness."

Earth had life.

V.

Spirit is a female component. Spirit, with her sub-component of lifetime, is life. Spirit is innately sinful, or the opposite of righteous, and remains sinful (without righteousness) while it is immature. Lifetime radiates sinfulness, or the opposite of righteousness, existing innately without righteousness. Spirit, as it is embodied, is light (Spiritual Light) until it become spiritual matter. In order to become spiritual matter, it must be nurtured by a mind righteously. Something is nurtured by something of a higher order (came before in time). Something cannot nurture itself.

In order to be nurtured by a mind and eventually become spiritual matter, spirit must be contained within a physical body in order that it be manifested by death light, bound within the three material dimensions. Earth is a first generation Spirit.

Earth was nurtured by both her father and mother. Her father, God, nurtured her spirit by radiating/shining His righteousness upon her. God nurturing matured Earth's "mind" (atmospheric righteousness). Her mother, the Sun, nurtured Earth's physical body. As Earth's physical body was nurtured came physical aspect to her atmosphere, along with the atmospheric righteousness.

Earth's life, as a result of nurturing, began to grow life, or living things. The Earth began to mature. The spirit of the living things that grew were an extension of her own spirit. The living things did not have spirits of their own. Through the life of the living things, Earth was more and better nourished by her atmosphere (the righteousness of God and the light and warmth of the Sun).

God impregnated Earth (Mother Earth) and Earth began conceiving another type of life:

Living beings. A living being, as opposed to just a living thing, is composed of three distinct and necessary aspects or components: A body (inherited from the material of Earth), a spirit (life, or spiritual light, with a sub-component of lifetime (time) inherited from the Father (God) and mind (which is an unmanifested body of light energy caused by the manifestation of lifetime within spirit). The composition of a living thing, as opposed to a living being, has only two of the three aspects or components: A body (extended from material of Earth) and a spirit (an extension of Earth's spirit).

Earth's spirit is nurtured righteously by the radiation of the mind of God. Spirit must be nurtured by a mind. Physically, she is nurtured by her mother, the Sun, which radiates more physical light (slower speed than mind light) and warmth. God's nurturing, by shining His high order righteousness upon the Earth, overpowers and balances the radiation of sinfulness of Earth's immature spirit. God nurtures Earth spiritually with also causing "atmospheric righteousness" and a manifestation of Earth's mind, though not within the three physical dimensions. It is not a physical manifestation. What makes it appear to be a physical manifestation is the radiation from the light of Earth's mother, Sun.

VI.

The conception process, the result of the impregnation, is a production of sound. The sound conceived must be manifested in order to create a body. Sound is manifested by light. Only light of some higher order (came before in time than the sound being manifested), radiated upon the sound, can manifest a body. There are only two orders of light that are higher than mind light and those are time light and death light, which is a higher order than time light.

Spirit must be nurtured by mind righteously in order to produce spiritual matter. Spirit is nurtured by radiation. Spirit is nurtured by the radiation of righteousness from the mind doing the nurturing.

Spirit, a female component, can also be impregnated by the mind – a male component. So the radiation (an order of light) nurtures the Spirit, and the mind doing the nurturing impregnates the spirit, producing spiritual matter. Spiritual matter is matter of another dimension that appears "like" light within the three material dimensions.

The spiritual matter of Earth, which is an order of sound and constitutes the "sound" of the bodies (living beings) she conceives, is not immediately manifested. Because matter of any order is some order of sound manifested by some order of light, spiritual matter can only be manifested by light of a higher order. Lights of higher order than spirit are time light and mind light. Time light is within spirit and a body cannot manifest itself. So the next higher order is mind light, God's light, as His light radiates or shines upon the Earth while she is being nurtured.

The spiritual matter of lesser beings is never manifested, except in the case of a human being whose soul has become divine. Where a human being's soul becomes divine, it's spiritual being (of which the spiritual matter is a part and is also of its lifetime and yet to be manifested mind body) is manifested by God's mind light.

Spirit is a component of a living being. It is a component but does not have a manifested body. Bodies are sound that are manifested by a high order of light. Spirit is a light body, with a sub-component of lifetime. Spirit is a light that is of such a frequency that it is also sound. Spiritual light is also spiritual sound. Spirit is the existence that is not one or the other, it is both. It is that existence of light and that existence of sound that is the intersection, and it is spirit and its cause: mind. Spirit is a light body when manifested prior to divinity. The light body becomes a sound body that can be manifested when it matures to a state of complete spiritual matter.

The light that manifests the spiritual light body is death light, until such time as the spirit's composition is mature and of spiritual matter (completely righteous), not just spiritual light, becoming divine. Then the spiritual body will be manifested by the mind light of God.

A manifested spiritual light body appears as the beings "aura" within the three physical dimensions. The appearance is the aura of the living being the spiritual light body is a component of.

Mind is similar to spirit in that it is light but it is light that exists of the light and time intersection: where light is time and time is light. It is an energy body that is sound (time sound) but it is not manifested. It is manifested simultaneously with the spirit (spiritual matter and lifetime) it nurtures once a marriage takes place and the spirit becomes divine. The marriage is of spiritual light body (with lifetime sub-component) and mind, provided the spirit and mind being married are contained within the same physical body. Mind does not appear at all within the three material dimensions, it is just an existence that can exist within the three material dimensions.

VII.

The body of God, a non-material, a body of energy, was manifested as it existed within a perfect vacuum. He exists in the nth dimensions of time, at the highest possible dimensions. Matter, sound and light moving at a much slower speeds, in order to be matter, must exist so far from God's speed, so much lower in the dimensions of time, that the dimensions of time within in which it exists become one and the same as the three physical dimensions of matter. The infinite vacuum is no longer perfect, because there exists matter, but the matter is so far away in time that the existence does not interfere greatly with God's environment.

VIII.

A nurtured spiritual body, whether a being or
thing which is nurtured with righteousness by a
mind, produces spiritual matter. Spiritual matter is
the production of spirit and mind. Spiritual matter
is the composition of the soul of the being or
thing, where the thing is a growth of the Earth.

The Earth has soul.

Earth possess the highest order of spirit. Spirits
lesser than Earth, her offspring or the descendants
of her offspring, are of Earth and are of a lesser
order.

The composition of the body of mankind is the
matter of Earth (part of her body and soul) with
the components of spirit (of Earth but lesser than
Earth) and mind (of spirit) of its own. Mankind's
body is first the housing for the immature (sinful)
spirit and his/her nurturing mind. Upon birth, the
mind is also immature. As his/her mind matures

and nurtures his/her spirit righteously, spiritual matter (righteous) is created. The spiritual matter created (now the soul) is contained within his/her body until death, when the spiritual matter exits.

If, during the lifetime, the spirit became divine, the divine spirit passes wholly as one Holy living spiritual being to its proper dimension. If the spirit did not become divine, that portion of the spirit that was spiritual matter (righteous) passes back to become a part of the soul of the Earth. The portion of the spirit that was still sinful dissipates and is suffocated by atmospheric righteousness.

A divine spirit is a spirit composed of completely righteous spiritual matter, which spiritual matter is married, or joined in Holy union, with the mind from which it received its nurturing. The two become one. In order to be granted an eternity of time, or an eternal lifetime, the spirit must become divine.

The spirit of Earth, nurtured by God, was the cause of the growth of living things on and of the Earth. The spiritual matter created by the beings and living things (which spirits are an extension of Earth's spirit and do not possess distinct spirits of their own) that exist on Earth, when not divine, pass back to the Earth, gradually to cause Earth's spirit, or soul, to be wholly not sinful, or righteously Holy.

Only the spirit of mankind has the ability to become divine and pass to the other dimension. This is because of the power and intellect of his/her mind as it is doing the nurturing. As far as lesser living beings and things that have produced spiritual matter during their lifetimes, never to become divine, upon their death the spiritual matter is passed back to Earth, becoming also part of Earth's righteous soul.

Man has the freedom of choice as to whether or not to nurture his/her spirit righteously, and create spiritual matter, or to feedback the innate sinfulness of his/her spirit. Where a spirit is not nurtured righteously, it will eventually die from lack of time (no eternity of time – no divinity) and not being anything physical of the spiritual dimension (no righteous spiritual matter) that is able to pass back to Earth.

The sinfulness left of a spirit that was either not or only partially nurtured with righteousness, upon the physical death of the body, is suffocated by the atmospheric righteousness that exists as a result of God's nurturing of the Earth. The spiritual matter (righteousness) that passes back to Earth upon the physical death of a non-divine body, is the soul of that non-divine body, returning to Earth.

It is the innate sinfulness of the spirit that provides the alternative to the thoughts of the innately righteous, but possibly immature, mind of a living being. The energy of the spirit is the lifetime. Lifetime is also the energy of the living beings mind. This connection permits a communication between the spirit and mind. Because the spirit's sinfulness is opposite mind righteousness, the mind can make a choice when an alternative to righteousness is presented.

IX.

In existence, we have the living being Earth, a
female celestial body, which was born of the
intercourse between her mother, Sun, and her
father, God. From her mother, Earth inherited the
physical aspects of her body. From her father, she
inherited lifetime. Lifetime, being an energy and
also being time, radiated light and created sound.
The light and sound were both of the same wave
and frequency. The light of the lifetime was also
the sound of the lifetime, but the sound passed
through a physical body or sub-body causing
mind. The result was a living spirit, the body of
light with lifetime, which was manifested by death
light, and mind (of Earth, manifested, of less than
Earth, unmanifested until divinity).

In her infancy, Earth began being nourished
physically by her mother and spiritually by her
father. The process of nourishing created an
atmosphere, Earth's atmosphere. Spiritual
nourishment came in the form of righteousness,

creating an atmospheric righteousness, which mind light of God penetrated Earth's body to nourish her spirit. As her spirit was nourished, it began to mature with spiritual matter being created.

During the maturation process, Earth started growing living things upon her body. These living thing had bodies, also, which were manifested by the mind light of the nourishment from God, but their spirits were an extension of Earth's spirit. Through these conduits, Earth was better able to receive her nourishment toward the creation of the spiritual matter of her soul.

Once blossoming, God had intercourse with Earth and Earth conceived living beings. These living things had bodies that could move about freely upon her surface, which were of Earth and manifested by atmospheric righteousness. These living beings had spirits of their own with the energy sub-component of lifetime, but the spiritual body that was a sub-body of the physical body was manifested by death light. Setting them apart from living things, these living beings also had the aspect of mind, which was the cause of the sound of lifetime as it passed through the physical body or sub-body of the being. Beings nourished their own spirits with their own minds.

Of these living beings were human beings, which were living beings of a high order. During the time they were alive, they had the opportunity to nourish their spirits with such righteousness that the spirit's innate sinfulness would be replaced with righteous spiritual matter. A marriage would take place between the spirit, with its composition of spiritual matter and sub-component of lifetime, and the mind. The lifetime would become eternal. The spirit would become a divine spiritual being and, upon the death of the physical body, the spiritual body would pass to another dimension: the dimension wherein the spiritual matter was physical matter. The divine spiritual body was manifested by the mind light of God.

If, during a lifetime, the living being did not or could not become divine, that portion of the spirit that was righteous spiritual matter (the being's soul) would exit from the corpse and pass back to Earth, becoming a part of the Earth's soul. That portion of the spirit that was still just light and sinful would dissipate and suffocate in the atmospheric righteousness.

X.

At this point, there are two ways the story could unfold. What direction to take would entail knowing exactly what God's objectives were regarding the creation. There could be the objective of "completion/advancement," or evolution. This objective would imply a definite point where the creation would be considered completed and it would pass on or graduate to another stage. It would evolve. The alternative is "perpetuity," which falls in with the reincarnation of souls.

The completion/advancement theory is that God would consider the creation complete and ready to advance when it became completely Holy. Indicative of this would be the spirit of the Earth being composed of all righteous spiritual matter (it will be Holy soul, not at all sinful). That spiritual matter had accumulated over time from constant nourishment and the pass back of the spiritual matter of living beings that did not become divine.

The souls, or spiritual matter, of non-divine beings would not be re-birthed, but remain a permanent part of Earth's soul with the completion/advancement theory. Human beings would have one lifetime to achieve divinity and become divine spiritual beings, else the opportunity would be lost forever as their souls would meld with Earth's soul.

Once Earth's soul was Holy, mankind would begin coming to an end as the divine spiritual beings would begin returning to Earth. The Earth and every being and thing upon it would pass to another dimension, where spiritual matter of the divine spiritual beings would become physical matter.

With the perpetuity theory, Earth's soul would never become Holy (completely righteous). As new beings were born, a portion of her righteous spiritual matter (part of her soul, which is composed also of the souls of deceased, non-divine beings) would be re-birthed into that new being. Human beings would be given lifetime after lifetime to achieve divinity. There would be no end to the reincarnations, until the human being became a divine spiritual being. The Earth would never graduate to the higher dimension and divine spiritual beings would remain in Heaven (or wherever, in another dimension) for the remainder of time.

A big problem with the reincarnation theory is "what if a being did not want to be reincarnated and did never want to become divine?" The reincarnation would not be a conscious decision to be made. It would just happen. It would be as if "freedom of choice" would be removed and as if saying that ALL human beings MUST eventually achieve divinity. If divinity were such a mandate, why is the path not taught in school or why is it not a standard way of living? Instead, the world is materially oriented and, it seems, if there were a direct route to divinity, the absolute way toward achievement is an unsolvable mystery.

For what would be the point of becoming divine, anyway? To sit in Heaven and never return home to Earth, where there could be a physical heaven to live out eternal life? What is there to do without a physical heaven, just play with the minds of living human beings? No physical happiness? What if the being didn't want to have to live again through the trauma of birth and then the frustrations of becoming mature to have to wait x number of more mortal years before continuing their quest for divinity?

I was more a follower of the completion/advancement theory. I was more inclined to think that divine spiritual beings will begin infiltrating the Earth at some point in time. More, I fathomed that a divine spiritual being would one day be able to keep his human physical body. It would be such that the physical body, not just the spiritual body, could achieve eternal life.

That would be an advancement. The Earth would pass to another dimension and, as it did, the divine spiritual beings with eternal physical bodies, that remain as one human being (divine), would somehow transform. They would not have to die and experience the trauma of leaving their human forms, for there is a "love" or attachment. They would be able to pass to the other dimension, a transformation would take place and they could greet the divine spiritual beings that were before as those beings became physical again.

XI.

Perpetuity is the secret to existence. It is the secret to energy. It is the secret to life. Perpetuity is something that once set in motion, would continue in motion forever, with no additional energy required to maintain it. It is an energy within itself. Divine Time is perpetual. Divine Mind is perpetual.

A perpetuity must be ignited by something outside itself in order to get the perpetual motion going. Impregnation is also the ignition of a perpetuity.

Divine Mind, or God, is the only perfect perpetuity. Should His energy, Divine Time, which is also perpetual but possibly not perfectly perpetual, ever cease its own perpetual motion, the motion is reignited by the mere thought of God's mind. God's thinking perpetually ignites the perpetuity of Divine Time, which energizes and causes God to be in existence perpetually.

There are imperfect perpetuities. An imperfect perpetuity is a perpetuity and seems to be perpetual but is only in perpetual motion (an energy within itself, but only apparently) for a certain amount of time. The Sun is an example of an imperfect perpetuity. It was ignited by God and is still perpetually burning in the Heavens. But what happens when an imperfect perpetuity's time runs out? It could be death or a transformation to another state or dimension: An evolution.

So the answer to the riddle is that the story continues perpetually but also evolves. There is both completion/advancement and perpetuity.

Imperfect perpetuity -> Evolution (transformation or death) -> New State of Imperfect perpetuity -> Evolution (transformation or death) … perpetually.

XII.

Time is an existence of the 4th dimension that exists within the three physical dimensions. It must be passed through. It is like there is perpetually a 4th dimension barrier after 4th dimension barrier that physical things or beings must pass through in order to exist or live for any period of time.

Whether or not a living being or thing physically transforms or dies at any point in time depends on how smoothly the being or thing passes through these time barriers. If the being or thing passes through the barrier, it transforms or continues. It ages. The smoother a being or thing transforms or continues determines the length of time the being or thing will live.

What causes death is physical trauma caused by erratic or chaotic transformations or continuances as the being or thing passes through the time barriers (passes through time). The body of the being or thing reaches the point where it is unable to handle any more trauma, it hits a barrier and literally gets its lights knocked out (it dies).

The key to living a long life, and possibly achieving physical eternal life, is to not cause oneself trauma, and to adopt a lifestyle that will carry the body through time smoothly and efficiently.

A human being is born with lifetime (an imperfect perpetuity of energy – time), mind (which, like God, could be abled to perpetually reignite the perpetuity of lifetime by its mere thinking- causing the human being to become a perfect perpetuity within himself), and spirit (which, with mind, is able to create spiritual matter).

Lifetime is time and therefore passes through time (the time barriers) as smoothly and efficiently as could be possible. Lifetime (being in motion) causes sound. The sound of lifetime is mind, which also moves through time as smoothly and efficiently as lifetime. Lifetime is what caused the light body of spirit. Light bodies also pass smoothly and efficiently through time. Spiritual matter will eventually become the physical body of the divine spiritual human being, melding with it the lifetime perpetuity and the mind.

Once lifetime and mind are physically connected, through its body of spiritual matter, mind can perpetually reignite the perpetuity of lifetime by the action of its thinking.

The speed of time's perpetual motion is far faster than the speed of light. Although lifetime (non-physical), mind (non-physical) and spirit (non-physical) pass through time easily, the motions and movements within a human body, and that cause the living of the human body (physical), are comparatively slow. A physical body's passing through time is not as naturally easy as the body's non-physical aspects passing through time.

Since the passage of a human being's lifetime (his time aspect) is most naturally smooth and efficient, in order to get the physical body to pass as smoothly and efficiently through time it must be in sync with its time energy. There must be a smooth pass of energy from lifetime to physical body and then from physical body to lifetime. There must be a perpetual back and forth that is not erratic or chaotic or trauma will result.

That is life and how life passes through time. It is a perpetual volley of energy between time (lifetime) and the energy of the physical body. The volley is not direct. It passes through mind (an unmanifested body of energy itself, to become manifested as one with lifetime and spirit upon achievement of divinity). Mind balances the volley and detains portions of the body's energy being passed back to lifetime in the event of an overload, releasing the energy gradually as to not overload lifetime. An overload to lifetime would cause death. Likewise, where the body energy constitutes and under-load, mind drains the physical body to make up the difference in order that the energy passing to lifetime is sufficient.

The human body gets its energy from nutrition…not necessarily physical food, eating of which typically causes the body to be more slow than normal – slower than necessary to pass smoothly and efficiently through time in sync with lifetime – and chaos or trauma being caused. Where there is chaos or trauma, the mind will work to balance it – but to continually traumatize the body will eventually cause death. The body will not be able to physically handle it.

It is more efficient for a body to obtain its nutrition in liquid form, eating physical food only in case of emergencies. An emergency could be considered as being called in a situation such as feeling light headed or passing out – the mind's signal to the body that its energy is way too fast and out of sync with lifetime and should be slowed down. The body should eat calories (physical energy) and then lie down or sleep until the energy is balanced.

Oppositely, if the body energy is moving way too slow, the mind will start to draw energy from the body and the individual will become tired or fall asleep.

If not slow or fast to the point of falling asleep or passing out, the next action of the mind would be to send the energy back for storage or draw upon the body's stored energy – the body will get fat or lose weight.

The consumption of nutrition should also be a smooth flow. For instance, a morning should begin with water – then move to coffee with milk – then move to milk or nutrition drink – then move back to coffee with milk – then move back to water. That would be a smooth intake of nutrition so as not to burst the body with energy by moving directly to the consumption of a nutrition drink. This smooth flow should be the example for all consumption of nutrition: work up to the burst (the bulk of the nutrition intake), then smoothly work out of it consuming other liquids until back to water.

Theoretically, then, if an individual can get his body energy moving in sync with his time energy (lifetime), so that there is a smooth volley of the energies, the physical body will simply flow through time without experiencing the trauma or constant trauma of crashing into a time barrier, leading to an eventual death. Eternal physical life is possible.

ABOUT THE AUTHOR:

Paeti Gustav Xaviers is the penname of Patti (Patricia J.) Markow, as she humbly efforts to record the thoughts, in as plain language as possible, that were inspired upon her by God. Her autobiography is written in the book *The Sound of Serenity*, also written under her penname.